Jian Marcel Zimmermann

Writers of the Cold

Jian Marcel Zimmermann

Writers of the Cold

Borges and Ramil on the Construction of Mythical Spaces

ScienciaScripts

Imprint

Any brand names and product names mentioned in this book are subject to trademark, brand or patent protection and are trademarks or registered trademarks of their respective holders. The use of brand names, product names, common names, trade names, product descriptions etc. even without a particular marking in this work is in no way to be construed to mean that such names may be regarded as unrestricted in respect of trademark and brand protection legislation and could thus be used by anyone.

Cover image: www.ingimage.com

This book is a translation from the original published under ISBN 978-613-9-66076-6.

Publisher:
Sciencia Scripts
is a trademark of
Dodo Books Indian Ocean Ltd. and OmniScriptum S.R.L publishing group

120 High Road, East Finchley, London, N2 9ED, United Kingdom
Str. Armeneasca 28/1, office 1, Chisinau MD-2012, Republic of Moldova, Europe
Printed at: see last page
ISBN: 978-620-7-98507-4

The poet whose name is linked to a place in the world is a modern and secure immortal.

Jorge Luis Borges

Everything is a dream
Everything is real
I never tire of living In Satolep Fields

Vitor Ramil

1

SUMMARY

This study analyses the literary construction of space. More specifically, we analyse the artistic construction of the city and the means used to give it high relevance in the textual sphere. To this end, we chose the work of Argentinian writer Jorge Luis Borges, published in the 1920s and 1930s: Fervor de Buenos Aires (1923); *Inquisiciones (1925); Luna de Enfrente* (1925); El *Tamano de mi Esperanza (1926); El Idioma de los Argentinos (1928); Cuaderno de San Martín* (1929); *Evaristc Carriego* (1930); *Discusión* (1932); *Historia Universal de la Infamia* (1935); *Historia de la Eternidad* (1936); which pay special attention to the city of Buenos Aires. Borges' work is of recognised aesthetic quality and has achieved an admirable capacity to influence generations to come. In this sense, we are analysing how Vitor Ramil (a self-confessed reader and admirer of Borges), who began his literary career in the 1990s, constructs his own Satolep from the city of Pelotas. For this analysis, his novels thematising the city are used: *Pequod* (1999) and *Satolep* (2008). Despite the distance in time that separates the two artists' writing, they write about a similar context, that of the early 20th century, defined by many theorists as *Modernity*, which is also the subject of our investigation. In order to achieve their aesthetic goals, the authors use different strategies, which we analyse individually, and then, in a relational analysis, we deduce the variants and constants in the creative process of the authors who form the analytical *corpus of* our study.

KEY WORDS: Literature. Modernity. Modernity. Jorge Luis Borges. Vitor Ramil.

CHAPTER 1

INTRODUCTION

This study aims to analyse an element that has always been relevant in literary production, but which is increasingly taking centre stage: literary "space", materialised here in the figure of the "city". To this end, we will use the work produced by Argentinian writer Jorge Luis Borges in the 1920s and 1930s, which is made up of the books Fervor de Buenos Aires (1923); *Inquisiciones (1925); Luna de Enfrente* (1925); El *Tamano de mi Esperanza (1926); El Idioma de los Argentinos (1928); Cuaderno de San Martín* (1929); *Evaristo Carriego* (1930); *Discusión* (1932); *Historia Universal de la Infamia* (1935); *Historia de la Eternidad* (1936). In these texts, great attention is paid to the writer's hometown, which he moulds in his own literary way. We will also make use of the work of a contemporary writer, the Brazilian Vitor Ramil, who in his works *Pequod* (1999) and *Satolep* (2008) creates, in an anagram of his city of origin, Pelotas, the fictional Satolep.

Jorge Luis Borges (1899-1986) is an icon of 20th century world literature. His vast oeuvre encompasses various textual genres, from poems, essays, narratives, "fictional essays", "fictional biographies", etc. His ingenuity makes it difficult to define his texts, which deliberately provoke the reader to think about literature as a whole. In general, he is defined as a universal writer, however, the basis for his production lies in his first books (according to himself), in which the city of Buenos Aires, especially the suburbs (las orillas), is given special attention.

Vitor Ramil has a curious career as a writer. When he released his first book in 1995, he was already a nationally known musician with a solid reputation, with four albums released and a good reception from audiences and critics. The author has a book of essays, *The Aesthetics of Cold: Geneva Conference* (2004), and three novels: *Pequod* (1995); *Satolep* (2008); *The Spring of Punctuation* (2014). For the purpose of analysing the literary construction of the city, we will only use *Pequod* and *Satolep*, in which there is a strong emphasis on this literary category.

We have brought these authors closer together for two reasons: firstly, because they use a similar theme, the city; and secondly, because of the influence that the Argentinian's work has on that of the Brazilian. The novel *Pequod* begins with an epigraph that is a quote from Borges: "La memória, esa forma del olvido" (RAMIL, 1999, p. 7), found in the poem El *ciego*, from the 1972 book El *Oro delos Tigres* (BORGES, 2015). In *Satolep*, there are many quotes from Borges, mainly from the poems *La Lluvia* (BORGES, 2015, P. 825) and *El Reloj De Arena* (BORGES, 2015, P. 815), such as the one below: "El tiempo: el curso irrevocable del agua que prosigue su camino" (RAMIL, 2008, p. 192), from the 1960 book *El Hacedor*. Ramil also declares himself a reader and admirer of Borges, and makes extensive use of the Argentinian's work in his musical production.

As expected, the cities created by the authors analysed focus on issues that go beyond their physical aspect, which is relevant as a basic building block in the compositions and allows for the literary creation of other aspects concerning more human issues, as can be seen in the subsequent analyses. In this sense, they both start from an objective city in order to subjectively create their *city*, not in the traditional ufanistic sense, but with virtues and problems that don't always coincide with factual cities. Both place their cities at the beginning of the 20th century, a time of many political, technological, warlike, sociological transformations, etc., a period many authors call "Modernity", which will also be the subject of our study, in order to obtain a more precise analysis of the structural intricacies of Buenos Aires and Satolep. Despite the similarities pointed out, there is an important difference, which may have implications for the way the text is produced: Borges deals with a Buenos Aires that is contemporary to the writing, while Ramil uses a Pelotas from almost a century ago as a base.

Buenos Aires at the beginning of the last century was one of the largest urban conglomerates in Latin America and one of the most advanced at the time (RAMA, 2001). It was also the "anchorage of astral fantasies" (SARLO, 2010) of modernism from Europe. Added to this was the general involvement with the celebrations of the centenary of Argentinian independence. In this context, Borges develops a unique *view of* the city, closes his eyes to the surrounding obviousness and produces, like many others, Argentine literature, but in his own way, with a focus on the shores of Buenos Aires, as authentic as the exalted Latin American metropolis.

The Pelotas used by Ramil in the creation of Satolep, on the other hand, is a city that is only in the imagination of its current inhabitants, given the time lapse between the context of the novel and that of its production. It is a developed and vibrant city, both financially (due to the massive production of jerky) and culturally, which, like most Latin American cities, also has social inequalities and prejudices of various kinds. However, Ramil doesn't create Satolep in order to criticise or exalt Pelotas, he creates it out of aesthetic necessity and artistic objectives, but without being oblivious to the virtues and imperfections of the original city.

Our study will minimally attempt to systematise the procedures adopted by the authors in the construction of their literary cities, in order to deepen the analysis into the hidden intricacies of the textual surface. Furthermore, we will intersect the strategies in order to clarify the phenomenon itself of creating mythical cities in literature.

CHAPTER 2

CITY/CIUDAD

> A description of Zaíra as it is today should contain all of Zaíra's past. But the city doesn't tell its past, it contains it like the lines of the hand, written on the angles of the streets, on the bars of the windows, on the handrails of the stairs...
> (CALVINO, 1990, p. 15)

Literature, both in prose fiction and poetry, operates on the basis of centuries-old categories and the more we are aware of how they work, the greater the possibility of working with them creatively and producing innovative variants. Our study focuses on analysing space in literary production, in order to see how it is constructed and the semantic possibilities developed there.

The space in which an artistic work is set can be of varying relevance, serving as a backdrop (sometimes minimally detectable) or even playing a leading role in the composition. This category is defined by Carlos Reis (2003):

> ...it comprises, in the first instance, the physical components that serve as a backdrop to the story: geographical settings, interiors, decorations, objects, etc.; in the second instance, the concept of space can be understood in a translational sense, encompassing both social atmospheres (social space) and psychological atmospheres (psychological space).... it is because of these choices that certain novelists are associated with the urban settings they prefer: if Eça is the novelist of Lisbon, Camilo is the novelist of Porto, Machado de Assis of Rio and Dickens of London. (REIS, 2003, p. 362)

More specifically, we'll look at the literary construction of a space: the city. In its categorical definitions, this space can be

defined as a demographic tangle made up of a considerable concentration of non-agricultural populations focused on commercial, industrial, financial and cultural activities (despite the countless contemporary variations). Pelbart describes the origin of cities in this way:

> ...the first recorded cities were meeting places to honour the dead, so that the cities of the dead precede the cities of the living... (PELBART, 2000, p.47)

Furthermore, the formation of the first cities is based on their function as "accumulators" of agricultural production from the surrounding area, acting as warehouses or greenhouses (mainly for cereals). According to Pelbart (2000), the city has the property of operating like a magnet, a magnetic field that attracts fragments of the most diverse natures and origins to its centre.

With regard to the formation of Latin American cities, their first function, according to Ángel Rama, was to dominate and civilise their surroundings, which was called 'evangelising" and later "educating", with a well-defined cultural basis:

> ...the cities were concrete applications of a general framework, the Baroque culture, which infiltrated the totality of social life and reached its culminating expression in the Spanish Monarchy (RAMA, 1998, p. 25).

What characterised the formation of these cities (Rama refers mainly to the cities of the Spanish colonies, but his definitions can be extended, with certain reservations, to the cities of the Portuguese colonies) was the attempt at a structural derivation of the city of Madrid. If we consider that Spain was already in economic decline at this time, we can see that Madrid was on the periphery of the European metropolises. In this sense, the cities founded in Latin America were the periphery of the periphery.

In these cities, there was a group of citizens responsible for "managing the feather", what Rama (1998) calls the "Literate City", which contrasted greatly with the "real city". This caste was made up of religious, administrators, educators, writers, intellectual servants... who were linked to the functions of power and determined a highly bureaucratic model of administrative functioning.

With regard to contemporaneity, Pelbart (2000) points to a process relatively similar to that of Latin American colonialism, in the sense of the subjection of cities:

> ...soon more than 80% of humanity will live in a single Megalopolis that makes all cities similar, abolishing the enormous diversity that made the glory and colour of the 16th century... the trend seems to be... the generic city. (PELBART, 2000,P. 46)

This configuration appears everywhere, especially in Asia, cities "without identity, without emblems, without a past, an orphan city, a city freed from the capture of the centre" (PELBART, 2000, p. 46), which remained after urban life migrated to cyberspace. What's more, in its current layout, first-world islands can be found anywhere on the globe, regardless of the surrounding third world and the misery that lies ahead.

According to Argentinian writer Ricardo Piglia (2005), art is a synthetic form of the universe, a microcosm that reproduces the specificity of the world. However, the transposition of the factual world into the literary one doesn't always take place in a linear fashion; through the bias of memory and imagination, this space can take on other colours:

> Man has imagined a city lost in memory and has repeated it as he remembers it. The
> real is not the object of representation but the space where a fantastic world takes
> place. (PIGLIA, 2005, P. 12)

The city, in this sense, is no longer reproduced as an objective image of thought, but as a product of the unconscious, with its superimposed layers, its evidence and ruins: if man inhabits a

(PELBART, 2000, p. 43).

It was this realisation that introduced the literary representation of this space more fully, to the extent that imagination and creativity, in a sense, became the "portrait" of cities, and writers were encouraged to produce it in their own way:

> When, from the end of the 19th century, the city was absorbed into the dioramas that
> unfolded symbolic languages and the whole of it seemed to become a forest of signs,
> its sacralisation by literature began. Poets, as the Cuban Julián del Casal said, are
> possessed of the "impure love of cities" and contribute to the arborescent
> *corpus* in which they are exalted (RAMA, 1998, p. 80).

Furthermore, again evoking an idea from Piglia (2005), if a map is a synthesis of reality, we need to be familiar with the place we now occupy on this map so that we can safely plot a future itinerary. In other words, in order to be effectively aware of the world we live in, we need to know both the "real" city and the "imagined" one.

According to Michel Foucault (2016), we live in what can be called "the epoch of space", in which our experience of the world is made up of a network that connects other people's points that intersect with its own skein. This is a context in which space is enriched and defined over time and, furthermore, sites become a form of relationship between various sites.

Furthermore, the author states that our spaces need to resolve certain conflicts, certain dichotomies:

> Our lives are still governed by certain insurmountable, inviolable dichotomies,
> dichotomies that our institutions have not yet had the courage to dispel. These
> dichotomies are oppositions that we take for granted: for example, between public
> space and private space, between family space and social space, between cultural
> space and useful space, between leisure space and work space. All these oppositions
> are maintained due to the hidden presence of the sacred (FOUCAULT, 2016, p. 2).

Gradually, however, these dichotomies seem to be diminishing, as one of the defining characteristics of our period is that it is iconoclastic (Bolanos, 2002), an era that questions and relativises its icons. Furthermore, the development of communication technology allows us to be virtually in several environments at the same time, dealing simultaneously with professional and personal issues, etc.

Foucault (2016) also draws attention to spaces that relate to other spaces, in a way that "second-guesses or inverts the network of relations that are themselves designated, mirrored and reflected. Spaces that are linked to each other, but contradict all the others" (FOUCAULT, 2016, p. 3).

These sites can be of two types: firstly, the author defines "utopias", which are spaces with no real place, that present a linear or inverted analogy with spaces that actually exist in society. These are unreal spaces that can present the world in an improved form or even turned upside down. There are also real spaces, "spaces that exist and are formed at the very foundation of society" (FOUCAULT, 2016, p. 3), which the author calls, in contrast to utopias, heterotopias. It's something like counter-sites, realised utopias, in which all the other real spaces of that given culture can be found, and where they are, at the same time, represented, contested and inverted.

What the author considers to be of great importance is that there can be spaces in between, where there is an intersection between utopia and heterotopia, which is a mixture similar to that of the mirror (since it is a place with nowhere, an unreal, virtual space that is open on the other side of the surface, a shadow that gives me visibility of myself).

Félix Guattari (1992) advocates the "restoration of the subjective city", in view of the way in which cities are being configured:

> The building and the city are types of object that, in fact, also have a subjective function. They are "objectivities" or, if you prefer, partial "subjectivities". These functions of partial subjectivation, which urban space presents to us, could not be abandoned to the taste of the property market, technocratic programming or the average consumer (GUATTARI, 1992, p. 178).

According to the author, in the near future, around 80% of the world's population will live in cities, and the rest will live in spaces related to urban centres, through various technological and civilisation links (health, education, economy, etc.). In this sense, it would be worth reflecting on the living conditions offered to these people, in view of the glaring contrasts that are currently being presented (in New York, one of the great centres of international finance, the streets of public parks are "invaded" by more than 300,000 homeless people).

Although cities are immense machines that produce individual and collective subjectivity, for the author, contemporary human beings are fundamentally deterritorialised:

> ...what could "their homelands" mean? Certainly not the place where their ancestors rest, where they were born and where they will have to die! They no longer have ancestors; they arose without knowing why and will disappear in the same way! They have some computerised numbers that are attached to them and that keep them under "house arrest" in a predetermined socio-professional trajectory, whether in a position of being exploited, assisted by the state or privileged. (GUATTARI, 1992,

p. 169)

In addition to an enormous flow of information and an unprecedented mobility, a large part of the population is forced to move to the big centres for very different reasons, be they economic, religious, warlike, political or emotional... In this way, there are two forces at work: one to homogenise cities more and more; and another that seeks to produce identity and differentiation niches from the heterogeneous formation of the urban population (which presupposes reflection and a critical attitude towards the condition of the "subjective city").

This is not just an attachment to citizens' original roots, but above all a democratic position of tolerance and freedom towards the organic hybridity inherent in the formation of cities, which the

he latent standardisation (whether conscious or a natural effect of our times) of contemporary urban lifestyles is rarely respected.

In The Poetics of Space, Bachelard (2016) reflects on the interior-exterior dialectic, indicating that this way of relating spaces denotes a way of rationalising the world:

> The deepest metaphysics is rooted in an implicit geometry, a geometry that - whether we like it or not - spatialises thought; if the metaphysician didn't draw, would he think? For him, the open and the closed are thoughts. The open and the closed are metaphors that he links to everything, including his systems (BACHELARD, 2016, p. 335).

In this sense, defining and reflecting on the space we occupy in the world (whether prosaically or literarily), contrasting it with the "outside", implies a binary view of the world, which "has the decisive clarity of the dialectic of *yes* and *no,* which decides everything. We make of this dialectic, without taking greater care, a basis for the images that command all thoughts of the positive and the negative". (BACHELARD, 2016, p. 335)

As much as "retiring to one's corner", reflecting on one's place of origin, is a trivial expression, and even a psychologically primitive image that generates simple images, for the author, "the simpler the image, the greater the dreams" (BACHELARD, 2016, p. 287). In other words, it is through critical reflection on our site that we develop a fuller vision of the world as a whole. Calvino (1990) materialises this idea in the following dialogue between the characters *Marco Polo* and *Kublai Khan*:

> - Every time I describe a city I say something about Venice.
> - When I ask about other cities, I want you to tell me about them. And about Venice when I ask you about Venice.
> - In order to distinguish the qualities of the other cities, I have to start from the first one, which remains implicit. In my case, it's Venice. (CALVINO, 1990, p. 82)

Given the contextual framework of the cities built in the works we will analyse in this study, it is important to pay attention to the

definitions of Modernity and the Modern City. Marshall Berman (2007) treats modernity as an

experience of time and space in which men and women from all over the world share the possibilities

and dangers of life. This is how the author defines this period:

> ...great discoveries in the physical sciences, changing our image of the universe and the place we occupy in it; the industrialisation of production, which transforms scientific knowledge into technology, creates new human environments and destroys old ones, speeds up the pace of life itself, generates new forms of corporate power and class struggle; a huge demographic explosion, which penalises millions of people uprooted from their ancestral habitat, pushing them along the paths of the world towards new lives; rapid and often catastrophic urban growth; mass communication systems, dynamic in their development, which wrap and bind the most varied individuals and societies in the same package; increasingly powerful national states, bureaucratically structured and managed, which stubbornly struggle to expand their power; mass social movements and nations, challenging their political or economic rulers, struggling to gain some control over their lives; (BERMAN, 2007, p. 16). 16)

Berman quotes Marx (2016) as saying that to be modern is to live in a world in which
"everything solid crumbles into air":

> All that is solid crumbles in the air, all that is sacred is profaned, and men are finally forced to face their real living conditions and their relationship with other men with more sober senses. (BERMAN, 2007, p. 86)

According to the author, there is no reason why every modern city should look like New York

or Los Angeles or Tokyo. However, we need to carefully evaluate the objectives of those who want

to armour their people against modernism, for their own "benefit".

The role that Literature can play, in the sense of de-alienating, forming critical awareness, in

the case of the structuring of cities, is to

played out effectively, for Berman, from Baudelaire's reflections[1] :

> Baudelaire shows us something that no writer has been able to see so clearly: how the modernisation of the city simultaneously inspires and forces the modernisation of its citizens' souls. (BERMAN, 2007, p. 142)

---

[1] Charles Pierre Baudelaire (Paris, 9 April 1821 - Paris, 31 August 1867) was a French poet and art theorist. He is considered one of the forerunners of symbolism and is internationally recognised as the founder of the *modern tradition in poetry.*

From this perspective, going back can be a way of moving forward, analysing the modernists of the 19th century can give us the vision and courage to create the modernists of the 21st century.

This lack of "solidity" can also be seen in Walter Benjamin's (2000) assessment of modernity. For him, cities have become as fragile as glass and, at the same time, as transparent as glass:

> The structure of Paris is fragile; surrounded by symbols of fragility. Natural symbols of creation - the black woman and the swan; and historical symbols - Andromache, "the widow of Hector and the wife of Helenus". Their common trait is sadness about the past and a lack of hope for the future. Ultimately, modernity approaches antiquity in this lapsed spirit (BENJAMIN, 2000, p. 17).

Important in this respect is the author's reflection on the role of the citizen (hero) in the city. To live in modernity would require heroic training, which does not presuppose a defined role: "Heroic modernity reveals itself as a tragedy in which the role of the hero is available." (BENJAMIN, 2000, p. 25). (BENJAMIN, 2000, p. 29) Available mainly to the "prosaic" man, since the wage labourer deserves the same applause and glory as the gladiator in antiquity.

Cruz (1994) also considers the study of the city to be relevant for understanding other constituent elements of a society:

> Thinking about literature through the lens of the city allows for a greater understanding of modern man and his conditions of existence, whether material or spiritual. Perhaps no genre is better suited to this task than the novel (CRUZ, 1994, p. 19).

Although we agree with the author, our corpus of analysis will be made up of novels and poems. Despite the structural differences, poetry and narrative can serve as raw material for analysing the city, the key is the literary capacity for expression and the aesthetic choices adopted, as the author himself later states:

> ...between this world view and the final text there are a number of factors that can make it more or less rich, more or less effective. Among them, mastery in the use of technical resources specific to literature (CRUZ, 1994, p. 43).

In short, both Borges' poems[2] and Ramil's novels will provide us with valuable material for a study that aims to analyse issues that move between literary aesthetics, sociological, intertextual and inter-media factors... operating, in short, on compositional frontiers, which can only be produced by authors with a sure skill.

---

[2] We are referring here to **Borges' "poems" because this is** the dominant **genre** in the author's books studied here. However, the texts of other genres present in these works will receive the same attention.

CHAPTER 3

BUENOS AIRES

Figure 1 (Xul Solar- Drago- 1927)

The writer Jorge Luis Borges is considered to be essentially universal, both for his aesthetic quality and for the wide-ranging themes of much of his literary output. However, at the beginning of his writing career, in the 1920s and 1930s, the author paid enormous attention to his place of origin, Argentina, and above all to the city of Buenos Aires, as Beatriz Sarlo points out:

> The city is not the content of a work, but its conceptual possibility. All the rural deviations in Rio de Janeiro literature of this century are produced in and from the city: people leave the city to write about the countryside. Literature visits the countryside, but lives in the city... Borges makes precisely the opposite movement: he imagines the city of the past with the language of a future literature. (SARLO, 1995, p. 8)

In this sense, we will evaluate the strategies used by the author in the literary construction of this space which, although it has solid references to the contemporary context of the writing, breaks into other paths that give the aesthetic object enough quality to guarantee its survival in the local and immediate conjuncture. To this end, the works Fervor de Buenos Aires (Fervour of Buenos Aires) will be used

(1923); *Inquisiciones (1925); Luna de Enfrente* (1925); El *Tamano de mi Esperanza (1926); El Idioma de los Argentinos (1928); Cuaderno de San Martín* (1929); *Evaristo Carriego* (1930); *Discusión* (1932); *Historia Universal de la Infamia* (1935); *Historia de la Eternidad* (1936).

The period we are analysing is immediately after the celebrations for the centenary of Argentinian independence, the May Revolution, which, as a rule, generates a certain ufanism in cultural production, which is not unrelated to the context in which it is produced, as Ángel Rama states:

> Literary works are not outside cultures, but crown them, and insofar as these cultures are secular and multitudinous inventions, they make the writer a producer who works with the works of countless men. A compiler, Roa Bastos would have said. The genius weaver in the vast historical workshop of American society (RAMA, 2001, p. 247).

The epigraph to this chapter features a canvas by a friend of Borges, Xul Solar, who, in addition to his friendship, shared with the writer a peculiar vision of Buenos Aires, based on the mixture of components from different places and times, in a basic hybridity:

> What Xul mixes in his paintings is also mixed in the culture of the intellectuals: European modernity and River Plate specificity, acceleration and anguish, traditionalism and a spirit of renewal, *criollismo* and the avant-garde. Buenos Aires: the great Latin American scene of a *mixed culture*. (SARLO, 2010, p. 32)

According to Sarlo (1995), Borges differentiated himself from the artistic/cultural production of the centenary, but he didn't give up on the national question. Although factually immersed in the country's modernisation, the author imagined a place still untouched by innovation and heterogeneity. However, this position does not imply a renunciation of his lineage, but rather a problematisation of the issue:

> There is no writer more Argentinian than Borges: like no other, he questioned the form of literature in one of the shores of the West. In Borges, the national tone doesn't depend on the representation of things, but on the presentation of a
>
> question: ^how can literature be written in a culturally peripheral nation? (SARLO, 1995, p. 3)

In the books that make up the analytical corpus of this work, we detected four different approaches to the city of Buenos Aires, which together form the mystique intended by the author: Physical City; Cultural City; Affective City; Everyday City. It is clear that we have developed this division for analysis purposes only, and this systematisation may not work for other proposals, but in this study it proves to be a useful tool. Furthermore, it is possible to see more than one of these approaches in the same text, as in the following example from the book Fervor de Buenos Aires:

Las Calles

The streets of Buenos Aires are already my entrance...
but the dull streets of the neighbourhood, almost invisible as usual, filled with
gloom and dusk and those further away
full of prickly trees
where austere cottages only venture out, surrounded by immortal distances, to lose
themselves in the honourable vision
of sky and moonlight.
A promise for the lonely
because thousands of unique souls pueblo them, unique before God and in time,
and undoubtedly precious.
To the West, North and South
The streets have been cleared - and so have the streets;
Ojalá en los versos que trazo
Estén estas banderas (BORGES, 2015, p. 17)

The poem above sometimes deals physically with the streets of Buenos Aires, but also affectionately states that they are *my neighbourhood...* In this simple example, we can see the heterogeneity of the author's treatment of the city, which together make up HIS city. Below, we will detail the approaches we have identified, stressing that this systematisation is not intended to be dogmatic, but only to act as a tool for understanding the author's complex construct.

Physical City

When Borges returned from Spain in 1921, Buenos Aires was in the throes of cultural upheaval, with the air of Modernism spreading, thousands of immigrants arriving, and changes of all kinds taking place at an accelerated pace. The city was changing, physically and culturally, which triggered in the author a certain aesthetic need:

Borges had to remember the forgotten of Buenos Aires at a time when that forgotten was beginning to disappear materially. This experience finds its poetic tone: the nostalgia of Fervor de Buenos Aires (SARLO, 1995, p. 9).

The author's poetry often tries to accurately describe certain places that are dear to him, as can be seen in the excerpt from the poem La Plaza san Martín, from the book Fervo de Buenos Aires:

All feeling is quiet under the absolution of the trees - jacarandas, acacias whose pious curves soften the rigidity of the impossible statue and in whose network the glory of the equidistant lights of the light blue and purple earth is exalted.
(BORGES, 2015, p. 21)

This poem appears in the author's first book, but it's not an ephemeral approach, because seven years later, in Evaristo Carriego, he continues to emphasise the physical, perhaps factual, accuracy of certain places in his hometown, as in the following excerpt:

> Palermo was an unconcerned poverty. La higuera oscurecía sobre el tapial; los balconcitos de modesto destino daban a días iguales; la perdida corneta del manisero exploraba el anochecer. On the humble walls of the houses, it wasn't uncommon to find some kind of manicured vase, aridly crowned with tunas: a sinister plant that in the universal sleep of others seems to correspond to a nightmare zone, but which is really suffered and lives in the most ungrateful terrains and in the deserted air, and which is distractedly considered an adornment. (BORGES, 2015, p. 108)

There are also countless other places in the city that are described in his texts, many of which are now considered tourist attractions (also in other works by the author produced at this time, but which have not been mentioned here). However, one space that receives special emphasis in Borges' work is the orillas, as defined by Sarlo:

> In those years, the term "orillas" used to refer to the remote and poor neighbourhoods, bordering the plain that surrounded the city. The orillero, a neighbour from these barrios, was often a worker in the slaughterhouses or meat-packing plants, where rural skills such as riding a horse and using a cart were still valued, is inscribed in a Creole tradition much more fully than the barrio compadrito (of whom Borges does not propose any idealisation), whose vulgarity denounces the newcomer or the imitator of customs that do not belong to him. The archetypal orillero descends from the Hispanic-Croatian tradition, and its origin is prior to immigration; the arrabalero compadrito, on the other hand, bears the marks of a low culture, and exaggerates the courage and challenge of the lighthouse to imitate the qualities that the orillero has as a nature. El compadrito es vistoso; el orillero es discreto y taciturno: (SARLO, 1995, p. 19)

The orillero predates the compadrito, and therefore comes from the era Borges revisits, in a mixture of nostalgia and the valorisation of authenticity, since the compadrito sometimes presents caricatures of the values and social practices of these suburbs. This space, devoid of glamour, becomes fertile literary material. At a time of celebration of Argentina's glorious achievements, Borges revered an Argentina that was less refined, but also worthy of note:

> Hacia el poniente quedaba la miseria gringa del barrio, su desnudez. The term "the shores" corresponds with supernatural precision to those shallow points, where the earth takes on the indeterminate nature of the sea and seems worthy of commenting on Shakespeare's insinuation: The earth has burbles, as does the water... Then: the Maldonado, dry and yellow, stretching out aimlessly from the Chacarita and which, by an astonishing miracle, went from the death of sed to the nonsensical expanses of violent water, which flowed with the dying rancherío of the shores. (BORGES, 2015, p. 109)

Borges presents man as a composite part of the environment, determining and determined by the physical conditions of the geographical space he inhabits, not as the traditional "fruit of the environment", so constant in certain literary periods, but the target of a reciprocal, living and changing influence.

Cultural City

Another of Borges' important concerns is the cultural manifestations developed in the city of Buenos Aires. However, this concern refers very little to literate culture; it focuses mainly on popular culture, in a mixture of documentation, imagination and aesthetic transformation. With regard to the book Evaristo Carriego, for example, which deals, roughly speaking, with the poet after whom the book is named, Sarlo (1995) states that it "pretends" to be a biography, when, in fact, it composes a Buenos Aires mythology. The author also states that

> Borges' literature, in the 1920s, emerges in this space of imagination. Like Xul Solar, he thinks that Buenos Aires needs aesthetic forms and strong cultural myths. However, unlike Xul Solar or Roberto Arlt, he first takes a tour of the 19th century and the Creole city: Borges travels. (SARLO, 1995, p. 9)

Borges himself talks about his motivations for writing Evaristo Carriego, which can perhaps also be applied to the other books that make up this research. In the author's own words:

> ^What was there, meanwhile, on the other side of the verge with spears? óWhat vernacular and violent destinies were fulfilled a few steps away from me, in the turbulent almacén or in the unlucky bucket? ^How was that Palermo or how beautiful was it? I wanted to answer these questions with this book, which is less documentary than imaginative (BORGES, 2015, p. 101).

Elements of popular culture are poetically detailed by the author, such as reactions to death, its symbolism, its rites (not by chance, there is more than one poem about the cemetery La

Recoleta, now a tourist attraction in the city). The games of truco, the peleas, the alcohol, form a kind of "home-grown mythology", exemplified in the following poem (extract from the poem El Paseo de Julio, from the book Cuaderno de San Martín):

> El Paseo de Julio
>
> I swear it's not deliberate that I've gone back to the high street of recovia repeated like a mirror, of barbecues with the meat of the Corrales, of prostitution concealed by the most distinctive thing: music...
>
> ...Eres la perdición fraguándose un mundo con los reflejos y las deformaciones de éste; sufres de caos, adoleces de irrealidad, te empenas en jugar con naipes raspados la vida; tu alcohol mueve peleas, tus griegas manosean envidiosos libros de magia... (BORGES, 2015, p. 95)

The author works with the city with such poetic power that the very elements that make up this space become confused with poetic elements: streets, houses, courtyards, are transmuted into stanzas, verses, etc. (extract from the poem Calle Desconocida, from the book Fervor de Buenos Aires):

> Maybe that time of the afternoon would give the street its tenderness, making it as real as a forgotten and recovered verse (BORGES, 2015, p. 20).

Furthermore, in this second approach to the city of Buenos Aires, the orillas once again receive special attention. The Argentinian saw himself represented more in the figure of the orillero than in that of the military man, his genius made it difficult for him to have close ties with the state (which differentiates him from North Americans or Europeans), his rebellious nature was identified with the man and the practices developed in the orillas:

> The orillas also derive from a found attribution. The arrabalero and the tango represent them. In a previous chapter, I wrote about how the arrabalero emerges from Corrientes Street and how the outpourings of El Canfaclaro, phonograph discs and radio, acclimatise this kind of actor in

> Avellaneda or in Coghlan. Its pedagogy is not easy: each new tango, written in the seditious popular language, is a hit, without missing the perplexing variants, the corollaries, the dark places and the reasoned disagreement of commentators. The joke is logical: the people don't need to join in local colour; the simulator says they do, but it's their hand in the operation. When it comes to music, tango isn't the natural sound of the neighbourhoods either; it was the sound of the burdels, in fact. The real representative is the milonga. Its current version is an infinite salute, a ceremonious gestation of rippling zalameras, corroborated by the deep strum of the guitar. Sometimes it narrates things of blood, time-honoured duels, deaths of valiant provocative charm; other times, it simulates the theme of destiny... The milonga is one of the great conversations of Buenos Aires; the truco is the other. (BORGES, 2015, p. 133)

This excerpt from Evaristo Carriego shows what kind of cultural manifestation is relevant to Borges, el arrabal offers what his imagination and perception require in the poetic construction of a sui generis literary space, as opposed to common sense, which saw only the wealth of Buenos Aires at the beginning of the 20th century.

Affective City

Despite the descriptive precision often employed, the rational tone, the inventiveness aimed at a specific purpose, eventually the author tempers his vision of the city of Buenos Aires with a certain affection. It's possible to say that, for a production that pays so much attention to a special geographical space, the author's affection for the place is at the conceptual basis of the work, acting as one of the motivators. However, sometimes this emphasis comes to the surface of the text, as in the poem La Vuelta, from the book Fervor de Buenos Aires, from which we quote an extract:

> La Vuelta
>
> After all the years of destruction, I have returned to the house of my childhood and
> I am still alien to its surroundings. My hands have touched the trees as if I were
> caressing someone who is suffering and I have repeated old paths as if I were
> recovering a forgotten verse and I have seen in the afternoon the fragile new moon
> rising from the shady shelter of the tall-leaved palm tree, like the bird in its flight...
> (BORGES, 2015, p. 36)

This affectivity exempts the author from producing a text with Cartesian precision, in which we can see a mixture of nostalgia, imagination, information and a certain purposeful "infidelity":

> Borges worked with all the meanings of the word "orillas" (margen, filo, límite,
> costa, playa) to construct an ideologeme that he defined in the 1920s and reappeared,
> until the end, in many of his stories. "Las orillas" are an imaginary space that stands
> as an unfaithful mirror to the modern city stripped of aesthetic and metaphysical
> qualities (SARLO, 1995, p. 19).

What's more, in some poems you can see that this affection is coloured by a melancholic/nostalgic tone. This is due to the mismatch that the author detects between his personal history and that of his ancestors. Borges doesn't share his grandparents' history of war and rural entrepreneurship, who became both soldiers and ranchers, capable of transforming a complex and hostile world into an almost bucolic environment: "y las cuatro estaciones fueran para ellos; como los cuatro versos de la copla esperada" (BORGES, 2015, p. 68). An example of this nostalgic melancholy can be seen in the poem Dulcia Linquimus Arva, from which we quote an extract:

> ...I'm a city dweller and I don't know about those things anymore, I'm a man
> of the city, of the neighbourhood, of the street: the neighbouring streets help me
> through my sadness with that long song they sigh in the afternoons. (BORGES,
> 2015, p. 68)

This fondness for his hometown is also reflected in the question of time, because it's not just a matter of passive nostalgia; the author believes that even his future is linked to the elements remembered/invented:

> This city that I believed to be my past is my future, my present;
> the years I lived in Europe are illusory, I was always (and will always be) in
> Buenos Aires.
> (BORGES, 2015, p. 32)

The mixture of tradition and imagination is a key feature of Borges' work from the 1920s and 1930s. In this sense, the aesthetic treatment devoted to this has a significant contribution from the author's affection for the themed city. After living in Europe for some time, the author began to realise the beauties and complexities of his homeland, a country that, as Sarlo (2010) states, lives in a "Peripheral Modernity":

> The first thing Borges does is invent a cultural tradition for this ex-centred place that is his country. This aesthetic and ideological operation runs through his work in the twenties and the first half of the thirties, up to Historia universal de la infamia, where he publishes his first short story (SARLO, 1995, p. 4).

The city, in short, becomes something vital for the author: "It is my story that Buenos Aires began: I consider it as eternal as water and air." (BORGES, 2015, p. 81)

Everyday City

To create his literary Buenos Aires, Borges uses a different approach: he uses everyday, trivial elements and places in order to give an account of the various aspects that make up this geographical space. There are poems about the rites of the end of the year (BORGES, 2015, p. 30), another about a butcher's shop (BORGES, 2015, p. 31), and one about an unknown street (BORGES, 2015, p. 31).

(BORGES, 2015, p. 20)... In the extract below, from the book Cuaderno de San Martín, this attention to the daily life of Buenos Aires is clear:

> North neighbourhood
>
> ...Once upon a time this neighbourhood was a friendship, an argument of
> aversions and affections, like other things of love;
> only if that faith persists
> in some distant events that will die:
> in the milonga that Cinco Esquinas is known for,
> in the patio like a firm rose under the growing walls, in the unpainted lettering that
> still says La Flor del Norte, in the guitar and baraja muchachos of the warehouse,
> in the detained memory of the blind man...
> (BORGES, 2015, p. 94)

Furthermore, as one of the author's favourite themes, the orillos are also represented in Borges' poetry. By incorporating the suburbs into the city, the orillero is not seen as an exception, but as an integral part of society:

> During the 1920s, Argentine literature took a turn that tended to focus on and develop a new point of view on marginalised people. The poor and marginalised become more visible as social subjects, the ways in which they are represented and the stories that are invented with them as characters change. (SARLO, 2010, p. 325)

However, valuing the orillas and orilleros does not imply idealisation. These elements are textualised with their virtues and defects, Borges presents the colour they give to the city and also the inconveniences they generate:

> Nineteen hundred sweet. In the many corralones of Calle Cervino or in the canaverales and huecos of Maldonado - an area dotted with zinc sheds, variously called salones, where the tango flamed, at ten centavos a piece and a companion - the neighbourhood would still tremble, and some man's face would turn historic, or a dead comrade with a human puncture in the belly would grow disdainful; but in general, Palermo conducted itself as God dictates, and it was a decent, unhappy thing, like any other gringo-coloured community. (BORGES, 2015, p. 130)

In this way, we can see that, for the purpose of the literary construction of Buenos Aires, Jorge Luis Borges pays attention to different aspects of the city, each with different approaches. Whether from a physical, cultural, affective or everyday perspective, what is clear is the author's understanding of the complexity of the formation of a geographical space, and that its heterogeneity also requires a multifaceted view in order to have a fuller and more secure analysis of this hybrid literary/social element.

CHAPTER 4

Pelotas/ `satoleP`

Vitor Ramil's hometown, Pelotas, serves as the basis for the space used in his fictional texts. Through an anagram, it is renamed Satolep, and other adaptations are also made in order to achieve the author's desired effect.

In general terms, we can say that Pelotas began to form at the end of the 18th century as a result of the installation of charqueadas, which had the ideal logistical conditions for such an endeavour. They took advantage of the abundant waterways to transport their production; the animals for slaughter were raised (or collected) in nearby estancias (for the parameters of the time); and the port of Rio Grande supplied the slaves needed to carry out the work.

The fortunes raised by the charqueadores were, without exaggeration, staggering, so much so that they made it possible to build mansions, palaces, castles? (which can still be visited today, and still amaze with their refinement and size) allowed their children to be educated in São Paulo, Bahia and even Europe. In addition, you have to consider the smaller fortunes built up in the orbit of the charqueadas, at the hands of merchants, industrialists...

This accumulation of capital led to the rapid development of the city, urbanisation above the Brazilian standards of the time and an active social life. In the second half of the 19th century, according to Magalhães (1993), with approximately 40,000 inhabitants, it was equal in this respect to Porto Alegre and São Paulo (Rio de Janeiro was far ahead). This figure becomes even more striking when you consider that most cities in France had around five thousand inhabitants.

The charque harvest took place between November and April, leaving the rest of the year idle for the charqueiros and many other residents of the city. It is mainly as a result of this factor that Pelotas develops numerous entertainment activities, many of them linked to culture. The nickname "Athens of Rio Grande do Sul" is due to the fact that the city has one of the oldest theatres in Brazil, the Sete de Abril (1834), which was constantly active in the 19th century, with Brazilian and foreign theatre companies and considerable local productions. Reading was also a regular activity among Pelotans at the time, further encouraged by the creation of the Pelotense Public Library (1875) which, in a luxurious building, made its vast collection available to readers. The abundant soirees also played an important role in shaping the city's culture, with interventions in other languages (mainly French) being not uncommon

According to Magalhães (1993), the peak of this development took place between 1860 and 1890. During this period, which followed the stagnation caused by the Farroupilha Revolution[3] , the

---

[3] We are talking here about the stagnation of development in the city of Pelotas, but we are not questioning the importance of the Revolution, which for ten years pitted the people of Rio Grande do Sul against the imperial

city of Pelotas did in fact have a nobility, mainly made up of Barons and Viscounts, who obtained these titles by freeing slaves and providing financial support for the war and economic issues of the Court.

Despite the elements that led the city to be known as the Princess of the South, there was also a scenario of exclusion, exploitation and a solid separation between social classes. What stands out most is the condition of slavery: much of Pelotas' wealth was built by the hand, blood and suffering of the slave labourers of the charqueadas. It is said that captives from other localities were threatened with being sent to the charqueadas in the south because of their bad behaviour.

Researcher Adão Monquelat has even published a book entitled Pelotas dos excluídos (MONQUELAT, 2015), based on newspaper reports from the city published in the 19th century, which denounce the difficulties experienced by those with limited purchasing power. In this work, there are many reports of precarious public health, sanitation, security and infrastructure, which is in stark contrast to the official historiography of the city.

In addition, the wealth of the charque brought with it prejudice, both racial and social class. This attitude can be seen in Magalhães' (1999) account of the Pelotas poet Lobo da Costa:

> Born in Pelotas, Lobo da Costa was an extremely bohemian spirit. In his verses, he
> often rebelled against the conservative austerity of the people of Pelotas. Even so,
> he never stopped loving his princess... (MAGALHÃES, 1999, p. 58)

In short, we can see the formation of a society, despite the era, that was plural and heterogeneous, capable of restricting certain soirees to "good people", but at the same time promoting the abolition of slavery four years before the Golden Law was passed.

In the period covered by Ramil's work, the beginning of the 20th century, the city of Pelotas had a different configuration from its prosperous past. The charqueadora activity had collapsed, for a variety of reasons, but we can list the main ones here: the emergence of meat packing plants, with the possibility of refrigerating meat instead of preserving it in salt; the absence of slave labour, which was also the main consumer of charque; the construction of the railway linking Bagé and Rio Grande, making this route more agile and profitable than the river flow through Pelotas.

In Rio Grande do Sul, during the Old Republic (18901930), there was also a geographical reorientation of the means of production, as well as a diversification of the products produced. With the decline of charque, the state began to have a polyculture economy, with an emphasis on the local market, shifting the centre of production to the Serra and Planalto regions of the state. The southern zone, where Pelotas is located, will experience economic stagnation, an example of which is that

---

administration, demanding a better balance between what was collected in taxes and what was returned as benefits to the state.

twenty new municipalities will be created in the state during this period, none of them in the southern zone.

In the cultural sphere, however, this crisis proved to be less serious. Even though, as a rule, cultural activities are closer to the economic centres, the "decadent" *Princesa do Sul* still had a considerable artistic scene. As an example, we can cite the emergence of epigones in the various arts: "In music, Zola Amaro; in plastic arts,

Leopoldo Gotuzzo; in literature, Simões Lopes Neto." (MAGALHÃES, 2012, p. 105)

According to Magalhães (2012), by virtue of tradition, Pelotas continues to flaunt its aristocratic and cultural image, but not that of opulence and wealth. Activities such as theatre, cinema, publishing houses, bookshops and higher education schools continue to proudly provide their services, but now to a society with less financial capacity.

Pequod

Vitor Ramil began his literary construction of Satolep in the novel Pequod (RAMIL, 1999), although in essays and also in his musical production, the characterisation of this space had been developed for years. In this initial construction, a very important aspect is the author's approximation between Satolep and the other cities of the Plata region, in this case, mainly Montevideo. This idea of homogeneity between the south of Brazil, Uruguay and Argentina is widely explained in A estética do Frio (Ramil, 2004), whose first appearances precede Pequod. The following extract exemplifies this procedure:

> Jugo y pulpa natural de naranjas VITAL, garantía de que las manhães de inverno en Uruguay eram todas las manhães de inverno de Satolep reunidas y aperfeiçoados de purza y calidad, su gusto y su claridad. (RAMIL, 1999, p. 59)

We can briefly link Ramil's claim to similarities to a variety of issues: the cold climate is characteristic of these places (which also distinguishes them from the rest of Brazil); the music produced in this region (in general) has many points of contact, as well as close origins; converging customs and habits (chimarrão; churrasco/assado...); economic activity based on the production of a similar matrix...

In the novel, one of the ways the cities are described is through the protagonist's father's habit of taking the family on car journeys around the city, especially to the outskirts:

> Ahab had brought us here in the same way that he used to take us on long, unusual night drives through remote, miserable villages with no pavements

or lighting. "This is Rua Francisco Lobo" or "There, Praça dos Enforcados" or "Look at all the plane trees in a place like this!", he would show us (RAMIL, 1999, p. 29).

In Uruguay, on a journey that father and son undertook in search of belongings left by his grandfather, who had lived in Montevideo, they travelled around the city by taxi, taking them to places that were dear to the father's heart and curious to the boy, who was also able to make connections between Satolep and Montevideo.

Another tool that Ramil uses in the intersection between spaces is the mixture of languages, which even appear in the same sentence:

> ...their advancement or their regression along the progressive path of culture. One of the most expressive was an oval, completely flat head, with no eyes, no mouth, no nose, nothing; (RAMIL, 1999, p. 15)

Despite the countless interpretations that such a strategy can suggest, the narrator himself explains his justification for mixing languages, without ending the matter or inhibiting the reader's different interpretations: "Spanish was then the language of thought and time. I was at the bottom of time.
'Eighty-five rooms and forty-nine baths' was enough for me." (RAMIL, 1999, p. 58)

Thus, in Pequod, Satolep is defined not by its singularity, but by the connection it establishes with other similar spaces. However, it is not a space without an identity, but an identity that is constructed by geographical, climatic and cultural issues, of an entire region that is very different from the rest of Brazil.

Satolep

The movement developed by Ramil's second novel takes the opposite direction to the first, in terms of the literary constitution of the city. If Pequod (RAMIL, 1999) strives for a relational definition with other spaces, Satolep (RAMIL, 2008) will focus on the singularisation of this specific place, and the fact that the novel is named after the fictional city already gives us an idea of its protagonism.

In order to create the theme city of the novel, we detected four approaches used by Ramil: the first is descriptive, using verbalised descriptions as well as factual photographic images[4] of the city

---

[4] These images were taken from the Pelotas Album (CARRICONDE, Clodomiro C. **Álbum** de **Pelotas 1922**. Historical Collection of the Bibliotheca Pública Pelotense).

of Pelotas at the beginning of the 20th century; the second is based on the Official History of Pelotas; the third uses the presence of local personalities from the most varied professional niches; the fourth is based on the author's internalisation of the previous approaches, producing an imaginary city, different from the factual Pelotas, but absolutely believable.

These divisions we have defined play an important role in analysing the configuration of Satolep's literary space. However, they are not developed linearly in the novel, sometimes they merge, and this interweaving brings homogeneity and fluidity to the text.

Description of Satolep

The text of the novel is structured as a statement by the protagonist, the photographer Selbor, given to a medical board that is judging his mental sanity. This statement is made up of a mixture of verbal text and photographs (more than 20, if you consider the enlargements) that set the scene for the recipients of the speech (whether the doctors in the plot or the external reader).

Ramil also uses the aforementioned device to physically establish the city of Satolep. Sometimes there is an independent verbal description:

> Once in the city, how beautiful... I came across a thriving industrial and commercial centre, which was also a countryside centre, since, because of the charqueadas, breeders, winterers and muleteers from all over the world converged here. The stone roads also opened up to the countryside, which I loved so much. In the midst of the development furore, education and culture flourished, as did the press, which I never abandoned and which never abandoned me. (RAMIL, 2008, p. 51)

On other occasions, the narrator, with the support of his profession, adds a photograph he produced to the verb, in order, according to Selbor himself, to provide material evidence for his story. Below is an example of the relationship between the photograph and its corresponding text:

(RAMIL, 2008, p. 18)

> In front of the Municipal Palace, a man is looking for a job, without realising that his shoes are pierced by perfectly polished cobblestones; a doctor fumbles with the memory of the dead patient in his hands, without directing his face to the statues, platbands and pediments that rise up around him; a smoker tied to a bagana breaks the oath of his last cigarette... (RAMIL, 2008, p. 19)

The relationship between photography and verbal discourse is a resource used by Ramil to define his Satolep. However, the sequence of verbal descriptions of the city and the corresponding image does not follow a linear pattern. The narrator's lacunar discourse potentiates other enigmas and, in this sense, sometimes the photographs introduced in the novel serve other purposes in the plot, instigating the reader to move through the countless mysteries that make up the plot.

Factual History

Another bias used by Ramil in the composition of his city is the use of the Official History of Pelotas. The formation of this urban agglomeration was, as already mentioned, due to the existence of the charqueadas and the wealth they produced. In this sense, it fits in with Pelbart's (2000) definition, in which cities are formed primarily to act as a warehouse for local production, and develop countless other activities that orbit around the matrix of production. Pelotas was structured in this way, and Ramil takes a critical approach to the subject, differing from the parochialism that is often perceived in historical discourse:

> It's April and the herds have disappeared from the Tablada. Do you smell carrion in the air? It rises from the cobblestones of the streets, from the floors of the shops. The charqueadas on the riverbank are just the visible face of a gigantic underground

27

slaughterhouse, the tips of an iceberg of slaughtered oxen that this tableau of streets
with beautiful houses covers up. (RAMIL, 2008, p. 127)

In principle, the allusion made in the quote above refers to the fact that the charqueadas'
activities extended from November to April. Furthermore, "carrion" can be interpreted
metaphorically, as a characteristic of a society that boasted unrivalled material wealth, but was built
on the foundation of prejudice (class, ethnicity, gender) and slave exploitation, as can be seen in the
excerpt below:

> "The slaves had to build the cave and then kneel down and pray to their
> master's saint," he explained. "As a form of resistance, they hid in it a small
> representation of themselves in their unworthy condition. By praying, they were
> asking Xangô for justice, which is what St John the Baptist represented for them.
> This city was built by slave labour. The thick walls, erected to protect the whites,
> hold the spirit of the blacks in their bricks." (RAMIL, 2008, p. 73)

Other, less controversial events from the history of Pelotas are also used in the novel as a form
of setting, and are seamlessly interwoven into the plot. For example, the flood that isolates Selbor in
a house in the middle of the countryside is a reference to the great flood that actually hit Pelotas in
1914.

Official History does not appear here as a backdrop (as it does in many traditional historical
novels), but with a dual function: to characterise the fictional setting; and also as a motto for critical
reflection on serious historical events, often obscured by the opulence (architectural, cultural...) that
endures to this day, as well as the evils imposed by the "nobility" on those without financial means.

Personalities from Pelotas

To construct Satolep, Ramil uses various personalities who existed in the city of Pelotas
(almost exclusively people linked to culture). In this sense, we can mention Menezes Paredes (1843-
1882), a poet who, despite some criticism, had a good reputation for his art. Francisco Santos (1864-
1937), a film-maker who many consider to be the forerunner of cinema in Brazil, who also becomes
an active character in the novel.

However, two personalities have different relevance to the plot.
The first is João Simões Lopes Neto (1865-1916), a writer whose best-known works are Lendas do
Sul and Contos Gauchescos (LOPES NETO, 1965), who lives directly with the protagonist Selbor
and is one of the people responsible for his staying in the city. In addition, in his aimless quest, Selbor
uses the legend of Negrinho do Pastoreio (present in Lendas do Sul) as a motto. Simões Lopes is so

important to the plot that, even after his death, he determines many of Selbor's passages and reflections.

Another essential character in the plot is the poet Lobo da Costa (1853-1888), a writer of unquestionable prestige in his time, but almost unmanageable in personal life, due to his alcoholism and temperament. He works with Selbor (and other friends) on the inauguration of the Guarany Theatre, and this relationship develops into a curious form of friendship. Finally, the death of Lobo da Costa is one of the decisive elements in the "mental confusion" that affects *Selbor and* leads *him* to be rounded up as a "street nut".

Despite the use of characters who existed extradiegetically, they make up a particular scenario, with many points of contact with the city of Pelotas, but many divergent ones. These characters are grouped together as contemporaries, who in fact were not. The most striking example of this process is the death of the characters João Simões and Lobo da Costa, in which there is an inversion of order, resulting in a time subversion of 28 years. In this way, Satolep's autonomy from Pelotas can be seen: the fictional city's setting is created on the basis of the real city, but it organises its nuances independently.

Furthermore, the use of personalities linked to culture leads us to consider Satolep, but especially Pelotas, as a place that fits Ángel Rama's (1998) definition of a "literate city". There are two points in which we can relate the concept to this space: firstly, because it was a society in which letters (meaning "the bureaucracy") had decisive power in the progress and development of this urban agglomeration; and secondly, although in a different context, the following words by Rama can be related to the space defined by Ramil:

> Although they were isolated from the hostile spatial and cultural immensity, it was up to the cities to dominate and civilise their surroundings, which was first called "evangelising" and then "educating". Although the first verb was conjugated by the religious spirit and the second by the secular and agnostic, it was the same endeavour to transculture from a European perspective (RAMA, 1998. p. 27).

Because of its countryside surroundings, it was Pelotas' responsibility to introduce a bit of "civilised" culture to its surroundings, which earned it the nickname "Athens of the South". We are aware of the limited scope of such a task, and of a certain naivety inherent in a thought that denotes a certain haughtiness, however, this is the image that prevails of the city of Pelotas in the 19th and early 20th centuries.

Imagined city? Ideal? Unreal?

Literary analysis can hardly do without an immaterial element and, in this sense, a very abstract definition: imagination. In the case of Satolep, where there are undeniable historical aspects in its construction, it is hard work to separate the factual and the imaginary, especially when it comes to space, which in this case still has an affective component:

> The house will simultaneously provide us with scattered images and a body of images. In either case, we will prove that imagination increases the values of reality. A kind of attraction concentrates images around the house (BACHELARD, 2016, p. 199).

The singularity of the story is already presented in the first line of the novel: "My visions of Satolep in ruins follow..." (RAMIL, 2008, p. 7) In other words, from now on the facets of a space that would also comprise other types of definitions will be presented subjectively, but the narrator traces two biases in the subjectivity/imagination/space relationship.

The first leads us to think of Satolep as a city that radiates images that stimulate transit between the real and the imaginary. The following words from the character Lobo da Costa indicate this direction:

> "We live in a city of mirrors. Or do you think that these giant suns and moons, these gloriously red clouds or these dreamlike twilights are what we are given to see? What guarantee do you have that the Satolep in which we live, immersed in this illusionist humidity, is not also an illusion? Ourselves: do you believe that we are who we see ourselves to be? I, a poet, say so!" (RAMIL, 2008, p. 221)

Geographical, cultural and architectural aspects operate this dichotomy between reality and illusion. The novel, however, does not present a solution to this "clash", nor does it assign a value judgement prioritising one of the aspects, it simply celebrates the amalgamation of aspects that are often seen as conflicting.

The second angle presented in the novel on the relationship between imagination and reality deals with another duality, which is also the motto of the narrator's discourse, between lucidity and insanity:

> BEING CRAZY IN SATOLEP. I ADMIT THAT IT'S A certificate of madness for someone to claim to be permeated by the voices and images of a city. But you can be sure that I will continue to open up to you sincerely, secure in the knowledge that, accompanied by material evidence, my account will appear to you to be reasonable and enlightening as a whole. (RAMIL, 2008, p. 246)

Selbor says he is "permeated" by the images of the city, both the images he sees and those that appear in photographs, and he also introjects the voices of this city, in other words, he assimilates the discourses of the city, about the city, both those that surround him and those of people who are dead or absent, but who often appear in his daydreams.

In the novel, these two paths establish the relationship between imagination and reality, which do not form a counterpoint, but rather a circle (an image dear to the narrator) that homogenises and makes the narrator's abstractions believable and fluent.

Finally, it is possible to realise that Ramil creates his Satolep from a complex web of strategies that sometimes highlight one element or another, but it is in the interweaving that they bring this hybrid composition together. From a real city, the author creates another, similar, but independent, enigmatic, surprising, an authentic literary city.

CHAPTER 5

INTERSECTIONS: BORGES- RAMIL

The influence that Borges' work has on Ramil's is evident, whether in the quotations made by the latter, his choice of themes, his approach to themes, etc. What's more, when it comes to the construction of fictional space, it's possible to see that they both start from the same origin: from their hometown, for which they have a peculiar affection, they develop their fictional spaces, with a series of similarities and differences to the original models. Both artists live for a period in other places and then return to their original cities (Borges lived in Europe; Ramil in Rio de Janeiro), which gives them a look at their origins in perspective. In this sense, it is undeniable that one of the motivations for the choices of Buenos Aires and Pelotas is affection. However, it is not a matter of ufanism, as it is a look that is also full of criticism, which directs the reader's attention both to the elements to be modified in these spaces, and to aspects present in these places that Official History and common sense purposely hide.

The compositional strategies of the two artists have many points of contact, and many of the elements present in the creative process are interconnected in such a way that their analytical dissociation becomes very complex. In the following words by Foucault (2016), when dealing more specifically with space, we realise this:

> In other words, we don't live in a kind of vacuum, in which individuals and things are placed, in a vacuum that can be filled with various shades of light. Rather, we live in a series of relationships that delineate sites that are decidedly irreducible to one another and that cannot be superimposed. (FOUCAULT, 2016, p. 3)

The naturalness of the interweaving, at times, even suggests that they are intrinsic. Many aspects of the category "city" harmonise literarily without conflict, aiming to create a general panorama, but without pretending to be totalising or dogmatic, geographically or ideologically.

Furthermore, we can see that the compositional basis of both authors is a similar process of transforming an objective image of the city into a subjective one. In the case of Borges, both the choice of the elements of the city that will be present in his literature and the treatment given to them contribute to this. By choosing the suburbs and their inhabitants (Orillas, Orilleros) as objects in principle, they end up becoming subjects who play a leading role in a Buenos Aires that does exist in reality, but in which their role is different, almost one of exclusion. Ramil's work, on the other hand, is more clearly perceptible in this process, as he starts from a photographic image, known to many as an "objective", and builds Satolep from it in his own way. Despite the impression of a totalising image, photography is also open to interpretation, as it both shows and conceals:

> Furthermore, for the purposes of analysing the photographic image, it is also necessary to consider the "out of field". These are the aspects that are outside the visual field of the image, only hinted at. Often, just as in verbal language, when what is not said, but implied, is of fundamental importance for understanding, in an image, what is purposely omitted or suggested is just as important as what is being shown (ZIMMERMANN, 2017).

Based on his reading of photographic images of Pelotas at the beginning of the 20th century, Vitor Ramil develops a literary city that is distinct from the real city. His aim is not to represent the actual city, but to create his own city.

Borges and Ramil ended up creating *Utopias,* in Foucault's definition: "Utopias are places without a real place. They are places that have a direct or inverted analogue relationship with the real space of society. They present society in a perfected form, or completely turned upside down." (FOUCAULT, 2016, p. 3) They start from a *Heterotopia*, which is a place that actually exists, and subvert some of its elements in order to develop their *Utopia*.

Furthermore, in both the Argentinian and Brazilian writer's conception, the city goes beyond its thematic and spatial role and becomes part of the linguistic structure of their texts. In the case of Borges, we can point to the aforementioned poem Calle Desconocida, from the book Fervor de Buenos Aires (BORGES, 2015, p. 20), in which the poet states that the street is as real as a verse. Vitor Ramil, on the other hand, intertwines the city with the verbal structure through the use of photographs of the city. These images do not have the classic function of illustration, they are integral and significant parts of the text, sowing information and mysteries, and "illustrating" very little. This strategy shared by both artists corroborates the protagonism of space in their works, making this literary category infiltrate the others, homogenising the hybrid textual construct.

Another point of contact between the authors in the formulation of their cities is the ambience produced by the music, whose genre also coincides, the Milonga[5] . In the Buenos Aires defined by Borges, despite common sense pointing to the Tango as a symbol of Buenos Aires, it is the Milonga that in fact represents the city, like an infinite greeting, corroborated by the serious sound of the guitar. He says that Tango is not the natural sound of the neighbourhoods, perhaps of the brothels. The milonga is also a fundamental genre for Ramil, both in his literary and musical work As well as having a disc called Ramilonga (Ramil, 1997), which has a pun between the author's name and the predominant rhythm on the disc, the Brazilian recorded a disc, Délibáb (RAMIL, 2010), made up exclusively of milongas (with lyrics by Jorge Luis Borges and João da Cunha Vargas[6]). In literature,

---

[5] Traditional style of music in Uruguay, Argentina and southern Brazil. Despite its many variants, milonga is treated here as a slow, repetitive, emotional rhythm, with a penchant for melancholy, density and reflection.
[6] It should be clarified that these are six lyrics by Borges, from the book **Para las Seis Cuerdas** (BORGES,

the novel Satolep features a character called Compositor, who sings milongas throughout the plot, providing a kind of soundtrack for the work. In addition, an important episode in the novel is named by the narrator as a "milonga night". Thus, it is possible to see the importance of an element that is different from the traditional literary structure, but which, in this case, contributes greatly to the formulation of the space, its image and its imagery.

Another coinciding strategy between the authors in the creation of their cities is the use of culture as a characteriser of space, but they do it in different ways. Although he was an intellectual scholar, it was the popular culture of the suburbs of Buenos Aires that Borges turned his attention to. In his texts we find references to wakes, festive dates, bars, brothels, cemeteries, etc.; which elevates to the category of protagonists practices, spaces and subjects that do not appear as such in the current imaginary of the city of Buenos Aires. Most of the characters in Satolep, a text in which Ramil pays special attention to the city, are artists, such as writers, filmmakers, actors, musicians and photographers. In this sense, culture is also an important tool for him in the construction of his fictional space. However, unlike Borges' choice, Ramil focuses on the production of literate culture, which, in a way, adds verisimilitude (although, in this case, dispensable) to the text, given that Pelotas, the city that serves as the inspiration for the novel, is renowned for its intellectual production.

In short, the influence of Borges' work on Ramil's artistic conception is undeniable, but from this ancestry he develops his own aesthetic standards. The authors share different tactics in their production, however, a more detailed analysis revealed that this is only superficially confirmed, that despite the similarities, each one, in their own way, traces the paths for their art to achieve its goal.

2015, p. 951), and six by João da Cunha Vargas, from the book **Deixando o Pago** (VARGAS, 1981), but put together by Ramil, without there being any interaction between the lyricists, both of whom had already passed away in 1997, when **Ramilonga** was released.

CHAPTER 6

FINAL CONSIDERATIONS

The phenomenon studied here can be defined, in Foucault's terms (FOUCAULT, 2016), as the creation of a Utopia from a Heterotopia. A fictional city is thus created from a space that already exists in the factual world. However, this new space is not reproduced as a linear and objective image, but with the fundamental collaboration of the unconscious, with its overlapping and mixed layers. These layers form a kind of network, which gradually coalesce to form a homogenous construct, a new space, but with a detectable origin. Furthermore, the sites also take shape from their relationship with other sites: Borges' Buenos Aires, orillera, is formed as a counterpoint to the glamourised metropolis; Satolep is built in relation to Montevideo, Buenos Aires and Pelotas itself.

It is also worth noting that we are dealing here with cities located in the modernist context of the early 20th century, which can be clearly related to Marshall Berman's definitions (BERMAN, 2007). For the author, this was an era characterised by great discoveries in the sciences; the industrialisation of production; the acceleration of the pace of life itself; the demographic explosion; the creation of mass communication systems... It is therefore configured as an experience of time and space, in which "everything that is solid falls apart in the air", producing a questioning of icons, the sacred, the immutable, which generates productive instability, previously unthought-of solutions to old problems and those arising from Modernity itself.

The city that Borges had chosen as the subject of his literature was at a time of change, both physical and cultural, which produced a mixture of European modernity and Rio-Platense specificity, acceleration and anguish, traditionalism and a spirit of renewal, creolisation and the avant-garde... This context led the author to question the form of Argentinian literature and the ways of producing art on the periphery of the West. In these early decades, his poetics worked to create a "mythology of Buenos Aires", with an eye on the past, present and future, producing a mixture of nostalgia and the avant-garde, without idealising any of the temporalities. In this sense, Borges' treatment of the city of Buenos Aires is heterogeneous, with cultural, physical, affective and everyday nuances that end up shaping his own city, neither factual nor idealised, but above all authentic.

From his hometown of Pelotas, Vitor Ramil creates the fictional Satolep, which appears in both his musical and literary output. In his first novel, Pequod, the artist characterises his fictional space based on its relationship with other spaces with similar characteristics: Satolep is shown in its similarity to Montevideo, which differentiates the southern Brazilian city from the rest of the country to which it belongs. This idea still appears in the author's second novel, *Satolep:* "We are on our way to expressing the transition between the Plata countries and Brazil, which is this place and which is us..." (RAMIL, 2008, p. 156), but the focus in this novel is different. Satolep operates in the characterisation and individualisation of this city, dealing with its conflicts and, above all, its people. The city now plays a leading role in the diegesis, and it is its uniqueness, both in terms of its faults and its virtues, that is emphasised by the narrator.

The author uses different approaches to construct his fictional city: he makes use of description (whether physical or in the abstract), the official history of the factual city that serves as the basis for his artistic creation, he uses local personalities as actual characters in the plot and, finally, he internalises these different aspects in order to interweave the construct.

The notorious influence exerted by Borges on Ramil's work, openly explained by the latter, propels him as a kind of aesthetic mainstay, from which the Brazilian develops new concepts and parameters for his artistic production, both musical and literary.

Lulled by the milonga soundtrack, both use their hometowns to create their fictional spaces. However, the literary city produced by Borges is in conflict with the stereotype of Buenos Aires, while Ramil produces Satolep without the purpose of duality, simply seeking to develop "his" city. In Foucault's terms (2016), both create Utopias from Heterotopias.

In this sense, local culture is a useful tool for both writers. Borges, in contrasting himself to the factual city, focuses on the practices developed in the suburbs of Buenos Aires (orillas), elevating popular culture and its subjects to the literary category. The cultural production of his hometown is also a central theme in Ramil's work, but with a different approach to that used by the Argentinian. The Brazilian builds his mosaic using exponents of literate culture, some of whom even become active characters in the diegesis.

Despite the affection that the authors obviously have for their native spaces, they thematise them in order to question common sense; they also aim to present the contradictions present in these spaces, the problems and virtues that structure these cities. However, both writers manage to avoid easy solutions that recur in the history of literature: their texts cannot be reduced to classification as social or political criticism, etc.; nor do they idealise the fictionalised spaces.

In short, both artists use a range of resources to make the "city" the protagonist of their works. Both do it in a very subjective way, without pretending to be dogmatic; they mix the real world with the literary, and from there they contribute to one of the most relevant and challenging activities: producing Art.

SUMMARY

This work analyses the literary construction of space as a protagonist. Specifically, we analyse the artistic construction of the city, and the means used to ensure that it has great relevance in the textual sphere. To this end, we chose the work of Argentinian writer Jorge Luis Borges, published in the 1920s and 1930s: Fervor de Buenos Aires (1923); Inquisiciones (1925); Luna de Enfrente (1925); El *Tamano de mi Esperanza (1926); El Idioma de los Argentinos (1928); Cuaderno de San Martín* (1929); *Evaristo Carriego* (1930); *Discusión* (1932); *Historia Universal de la Infamia* (1935);

Historia de la Eternidad (1936); which pay special attention to the city cf Buenos Aires. Borges' work is of recognised aesthetic quality, and has developed a great capacity to influence subsequent generations. Therefore, we invest igamos de qué modo Vitor Ramil (lector y admirador de Borges), que empezó su carrera literaria en los anos '90, construye, desde la ciudad de Pelotas, suia Satolep. For this analysis, we used his novels that deal with the city: *Pequod* (1999) and *Satolep* (2008). Despite the distance in time that separates the literary creation of the two artists, they write about the same context, the beginning of the 20th century, defined by many theorists as *Modernity*, which is also the subject of our research. In order to achieve their aesthetic goals, the authors use multiple strategies, which we delve into individually and then, in a relational analysis, glimpse the variables and constants in the creative process of the authors who form part of the analytical *corpus of* our work.

KEYWORDS: Literature. City. Modernity. Jorge Luis Borges. Vitor Ramil.

CHAPTER 7

REFERENCES

AGAMBEN, Giorgio. Stanzas: the word and the ghost in Western culture. Translated by Selvino José Assmann. Selvino José Assmann. Belo Horizonte: Ed. UFMG, 2007.

ARAGON, Louis. The Peasant of Paris. Translated by Flávia Cristina Souza Nascimento. Available at   : <http://www.bibliotecadigital.unicamp.br/document/?code=vtls000035077>. Accessed on 15 August 2016.

BACHELARD, Gaston. The poetics of space. Available at: <https://bibliotecadafilo.files.wordpress.com/2013/11/bachelard-a-poc3a9tica-do-espaco.pdf> Accessed on 05 Sep. 2016.

BAUDELAIRE, Charles. The painter of modern life. Available at: <https://disciplinas.stoa.usp.br/pluginfile.php/7789/mod_resource/content/1/BA UDELAIRE_O%20pintor%20da%20vida%20moderna.pdf>. Accessed on 12 September 2016.

BAUMAN, Zygmunt. Liquid modernity. Translated by Plínio Dentzien. Plínio Dentzien. Rio de Janeiro: Jorge Zahar, 2001.

BENJAMIN, Walter. Modernity and the Moderns. Trad. Heindrun Krieger Mendes da Silva Arlete de Brito; Tânia Jatobá. Rio de Janeiro: Ed. Tempo Brasileiro, 2000.

BERMAN, Marshall. All that is solid vanishes into thin air: the adventure of modernity. Trad. Carlos Felipe Moisés; Ana Maria l. Ioriatti. São Paulo: Ed. Companhia das Letras, 2007.

BOLANOS, Aimée G. Pensar La Narrativa. Rio Grande: FURG, 2002.

BORGES, Jorge Luis. Complete Works. Available at: <https://literaturaargentina1unrn.files.wordpress.com/2012/04/borges-jorge-luis-obras-completas.pdf>. Accessed on: 25 August 2015.

. *Obras completas 1. 2ª* ed. Buenos Aires: Sudamericana, 2016.

CALVINO, Italo. *Invisible Cities*. Translated by Diogo Mainardi. Diogo Mainardi. São Paulo: Ed. Companhia das Letras, 1990.

CARRICONDE, Clodomiro C. *Album of Pelotas 1922*. Historical Collection of the Bibliotheca Pública Pelotense.

CESAR, Guilhermino. *History of the literature of Rio Grande do Sul*. Porto Alegre: Ed. Globo, 1971.

CORRÊA, Gilnei Oleiro. *The city, the armchair and the line: studies on the aesthetics of cold, by VitorRamil*. Disponível em: http://antares.ucpel.tche.br/poslet/dissertacoes/index.php?subdir=mestrado%2f 2013&sortby=name. Accessed on: 01 July 2015

______________________ . *Redes de gelo: Estudos sobre a estética do frio, by Vitor Ramil*. Pelotas, Ed. Gráfica Universitária, 2009.

CRUZ, Cláudio Celso Alano da. *The invention of the (peripheral) neighbourhood in Buenos Aires literature: the case of Evaristo Carriego*. In: CRUZ, Claudio (Org.), *Outra Travessia* / Federal University of Santa Catarina. Centre for Communication and Expression. Postgraduate Programme in Literature - no. 8 (2nd semester, 2009). Florianópolis: EDUFSC, 2009.

______________________ . *Literature and the modern city: Porto Alegre 1935* Porto Alegre: Edipucrs: IEL, 1994.

ESTRADA, Ezequiel Martínez. *Radiography of the Pampa*. Rio de Janeiro: UFRJ, 1996.

Figure 1. SOLAR, Xul. Drago. Available at :< http://www.minutouno.com/notas/18971-xul-solar-drago>. Accessed 21 Mar. 2017.

FOUCAULT, Michel. Of OtherSpaces . Available at:< http://links.jstor.org/sici?sici=0300-7162%28198621%2916%3A1%3C22%3AOOS%3E2.0.CO%3B2-F>. Accessed on 26 July 2016.

GUATTARI, Félix. Chaosmosis: A new aesthetic paradigm. Translated by Ana Lúcia de Oliveira and Lúcia Cláudia Leão. Ana Lúcia de Oliveira and Lúcia Cláudia Leão. São Paulo: Editora 34, 1992.

HAESBAERT, Rogério; Marcos Mondardo. Transterritoriality and anthropophagy: transit territorialities from a Brazilian-Latin American perspective. Available at :< http://www.uff.br/geographia/ojs/index.php/geographia/article/viewArticle/378>. Accessed 26 July 2016.

LOPES NETO, João Simões. *Gaucho tales and legends from the south.* 3ª ed. Porto Alegre : Globo, 1965.

MAGALHÃES, Mario Osorio. *History and traditions of the city of Pelotas.* Pelotas: Ed. Armazém Literário, 1999.
Opulence and culture in the Province of São Pedro do Rio Grande do Sul: a study of the history of Pelotas. Pelotas: EdUFPEL: Co-edited by Livraria Mundial, 1993.

________________________________. *Pelotas princess.* Pelotas: Ed. Diário Popular, 2012.

MARX, Karl. *Manifesto of the Communist Party.* Available at: <http://www.pcp.pt/sites/default/files/documentos/1997_manifesto_partido_com unista_editorial_avante.pdf>. Accessed on 24 November 2016.

MASSEY, Doreen. Politics and Space/Time . Available at: < https://newleftreview.org/I/196/doreen-massey-politics-and-space-time>. Accessed on 26 July 2016.

MONQUELAT, Adão F. Pelotas dos excluídos (subsidies for a history of everyday life). Pelotas: Ed. Livraria Mundial, 2015.

________________________ . If Pelotas did not create charque, charque and
________________________ slavery invented Pelotas .
at:<http://pelotasdeontem.blogspot.com.br/2016/07/se-pelotas-nao-criou-o- charque-o.html>. Accessed on 08 August 2017.
Invitation to Reflection - Pelotas Anniversary Chaplaincy and Christian Identity . Available at:<http://pelotasdeontem.blogspot.com.br/2016/07/entrevista-de-adao- monquelat-concedida.html>. Accessed on 08 August 2017.

PALERMO, Zulma. Critical discourse in Latin America II. Buenos Aires: Corregidor, 1999.

PELBART, Peter Pál. Vertigo by a Thread: Politics of Contemporary Subjectivity. São Paulo: Ed. Iluminuras, 2000.

PIGLIA, Ricardo. The last reader. Barcelona: Ed. Anagrama, 2005.
_______________ . The absent city. Barcelona: Ed. Anagrama, 2006.

RAMA, Ángel. La ciudad letrada. Montevideo: Ed. Arca, 1998.
_______________ . Literature and culture in Latin America. São Paulo: Edusp, 2001.

RAMIL, Vitor. *Ramilonga: The aesthetics of cold.* Rio de Janeiro: Satolep Discos; Estúdio CIA dos Técnicos, 1997.

Pequod. Porto Alegre: L&PM, 1999.

. *Aesthetics of the Cold: Geneva conference.* Pelotas: Satolep Books, 2004.

. *Satolep.* São Paulo: Cosac Naify, 2008.

. *Délibab.* Buenos Aires: Satolep Music; Circo Beat, 2010.

_______. *Neutral Fields.* Porto Alegre: Satolep Music; Estúdio Áudio Porto, 2017.

_______. *Neutral fields: songbook/Vitor Ramil.* Pelotas: V. Ramil, 2017.

REIS, Carlos. *The knowledge of literature: introduction to literary studies.* Porto Alegre: EDIPUCRS, 2003.

RUBIRA, Luis. *Vitor Ramil- Nascer leva tempo: identidade, autossuperação e criação de Estrela, estrela a Longes.* Porto Alegre: Pubblicato Editora, 2015.

SARLO, Beatriz. *Peripheral modernity: Buenos Aires 1920s and 1930s.* Translated by Júlio Pimentel Pinto. Júlio Pimentel Pinto. São Paulo: Cosac Naify, 2010.

_______________ . *Borges, un escritor de las orillas.* Available at:
<http://lproweb.procempa.com.br/pmpa/prefpoa/festinverno/usu_doc/6761331- sarlo-beatriz-borges-un-escritor-en-las-orillas.pdf>. Accessed on 06 September 2016.

SENNETT, Richard. *Flesh and stone: the body and the city in Western civilisation.* Trad. Marcos Aarão Reis. Rio de Janeiro: Ed. Record, 2003.

SPIVAK, Gayatri Chakravorty. *Can the subaltern speak?* Belo Horizonte: UFMG Publishing House, 2010.

STAUDT. Sheila Katiane. *Urban portraits in Brazilian novels of the 21st century: a reading of "they were many horses", "the photographer" and "satolep".* Available at:< https://www.lume.ufrgs.br/bitstream/handle/10183/131536/000973665.pdf?sequ ence=1>. Accessed on 03 September 2016.

STROUD, Christopher; Dmitri Jegels. Semiotic landscapes and mobile narrations of place: Performing the local. Available at:< https://www.tilburguniversity.edu/upload/efbfea1e-3d45-44cf-9f76- ae9fdccb8f23_TPCS_50_Stroud-Jegels.pdf>. Accessed on 26 July 2016.

VARGAS, João da Cunha. Leaving the paid. Porto Alegre: Habitasul, 1981.

ZIMMERMANN, Jian Marcel. From the representation of the image to the image as representation: Interdisciplinary relations in Satolep and The Portrait Painter. Saarbrucken, Deutschland: Ed. OmniScriptum. 2017.

CHAPTER 8

ANNEXES

INTERVIEW WITH MRS SARA SARTORE, EMERITUS COLLABORATOR OF THE JORGE LUIS BORGES INTERNATIONAL FOUNDATION

Jian: ^Is the public that most visits the Foundation made up of Argentinians or foreigners?

Sara: Actually, the place where Borges' work is most valued and appreciated is abroad. Here he was criticised a bit, another part that, of course, wanted him very much. It hasn't been understood, perhaps, so it's not valued as much. They come from abroad, even from Russia, from Japan, from China, Borges' work is very widespread in China, and also, of course, from Europe. All those countries that he mentions in his literature, but here in Argentina, unfortunately, he doesn't mention Borges' work.

Jian: ^Las escuelas, los ninos, visitan la fundación, leen a Borges?

Sara: Well, people are reading a bit more, there's more interest. The foundation, twenty years ago, started with a poetry competition, and that awakened attention to Borges' work. It's aimed at high schools, at secondary school students, but also at teachers, who have begun to deepen their understanding, and to see, and are increasingly incorporating readings of Borges' stories and poetry. And the root of all this is the poetry competition, which was started by Mrs Codama. And this has also opened up panoramas for young people, and this interest in reading Borges, how to approach him, because many say "no, it's very difficult", but if we start reading the essays, yes, it is difficult, but if we approach the stories, and the poems, which are part of an emotion, part of an experience. But this has also opened up a wide range of studies by teachers specialising in other areas.

Jian: Buenos Aires is a city where there are many protests. Is Borges' name there at these times, like a symbol or something?

Sara: The protests go the other way. Of something, sometimes invented, sometimes fattened up.

Jian: ^How does Borges' relationship with the city seem to you?

Sara: The thing is that Borges was a great walker, he walked all over the city, and experienced all the emotions of that place, and that's why he writes, whether about Palermo or La Boca, he also has other writings about Plaza San Martin. When you get to know these places, you understand them. So he was a great walker, a connoisseur of Buenos Aires, he loved Buenos Aires, he really felt very porteño, he didn't say porteño, he said Buenos Aires. Many people criticise him here, saying that he was an outsider, but that's not true.

Jian: J_a ciudad tradicional, con tangos, milongas... es solo para los turistas o aún sobrevive?

Sara: If you can see beyond what's in front of you, you can survive.

IMAGES OF BUENOS AIRES[7]

Plaza San Martín, a space themed by Jorge Luis Borges' poem Homonymous, analysed in this study.

---

[7] The photographs of Buenos Aires presented here were taken by the author in November 2017.

Plaza San Martín

Plaza San Martín

Recoleta Cemetery, an area of Buenos Aires dear to Borges, the subject and setting of many of his texts.

Recoleta Cemetery.

Recoleta Cemetery.
Jorge Luis Borges in the city of Buenos Aires

Borges Cultural Centre

Borges Cultural Centre

Jorge Luis Borges International Foundation and Borges Museum

Jorge Luis Borges International Foundation and Borges Museum in "The Night of the Museums", 04 Nov. 2017.

The first image appears in the novel Satolep (RAMIL, 2008, p. 18); the second photograph was taken by the author of this study in November 2017.

Both refer to the same location, Pelotas City Hall, and the time lapse between them is approximately 96 years.

More
Books!